Managing the Moon Program

Lessons Learned From Project Apollo

Proceedings of an Oral History Workshop
Conducted July 21, 1989

Moderator:
John M. Logsdon

Participants:
Howard W. Tindall
George E. Mueller
Owen W. Morris
Maxime A. Faget
Robert A. Gilruth
Christopher C. Kraft

MONOGRAPHS IN AEROSPACE HISTORY
Number 14
July 1999

National Aeronautics and Space Administration
NASA History Division
Office of Policy and Plans
Washington, DC 20546

Foreword

In a spring 1999 poll of opinion leaders sponsored by leading news organizations in the United States, the 100 most significant events of the 20th century were ranked. The Moon landing was a very close second to the splitting of the atom and its use during World War II. "It was agonizing," CNN anchor and senior correspondent Judy Woodruff said of the selection process. Probably, historian Arthur M. Schlesinger, Jr., best summarized the position of a large number of individuals polled. "The one thing for which this century will be remembered 500 years from now was: This was the century when we began the exploration of space." He noted that Project Apollo gave many a sense of infinite potential. "People always say: If we could land on the Moon, we can do anything," said Maria Elena Salinas, co-anchor at Miami-based Spanish-language cable network Univision, who also made it her first choice.

With his 81-year-old eyes, historian Schlesinger looked forward toward a positive future and that prompted him to rank the lunar landing first. "I put DNA and penicillin and the computer and the microchip in the first 10 because they've transformed civilization. Wars vanish," Schlesinger said, and many people today cannot even recall when the Civil War took place. "Pearl Harbor will be as remote as the War of the Roses," he said, referring to the English civil war of the 15th century. And there's no need to get hung up on the ranking, he added. "The order is essentially very artificial and fictitious," he said. "It's very hard to decide the atomic bomb is more important than getting on the Moon."

There have been many detailed historical studies of the process of deciding on and executing the Apollo lunar landing during the 1960s and early 1970s. From the announcement of President John F. Kennedy on May 25, 1961, of his decision to land an American on the Moon by the end of the decade, through the first lunar landing on July 20, 1969, on to the last of six successful Moon landings with Apollo 17 in December 1972, NASA carried out Project Apollo with enthusiasm and aplomb.

Of all the difficulties facing NASA in its bid to send humans to the Moon in the Apollo program, management was perhaps the greatest challenge. James Webb, NASA administrator from 1961 to 1968, often stated that while the technological aspects of reaching the Moon were daunting, these challenges were all within grasp. More difficult was ensuring that those technical skills were properly utilized and managed. Thus, the success or failure of Apollo in large part depended on the quality of its management. "We can lick gravity, but sometimes the paperwork is overwhelming," Wernher von Braun once said.

To a very real extent, Project Apollo was a triumph of management in meeting enormously difficult systems engineering and technological integration requirements. NASA leaders had to acquire and organize unprecedented resources to accomplish the task at hand. From both a political and technological perspective, management was critical. The technological accomplishments of Apollo were indeed spectacular. However, it may be that the most lasting legacy of Apollo was human: an improved understanding of how to plan, coordinate, and monitor the myriad technical activities that were the building blocks of Apollo.

More to the point, NASA personnel employed a "program management" concept that centralized authority over design, engineering, procurement, testing, construction, manufacturing, spare parts, logistics, training, and operations. The management of the program was recognized as critical to Apollo's success in November 1968, when *Science* magazine, the publication of the American Association for the Advancement of Science, observed:

In terms of numbers of dollars or of men, NASA has not been our largest national undertaking, but in terms of complexity, rate of growth, and technological sophistication it has been unique....It may turn out that [the space program's] most valuable spin-off of all will be human rather than technological: better knowledge of how to plan, coordinate, and monitor the multitudinous and varied activities of the organizations required to accomplish great social undertakings.

The editor of *Science* probably did not fully understand the complex project management procedures used on Project Apollo.

While there have been many studies recounting the history of Apollo, at the time of the 30th anniversary of the first lunar landing by Apollo 11, it seems appropriate to revisit the process of large-scale technological management as it related to the lunar mission. Consequently, the NASA History Office has chosen to publish this monograph containing the recollections of key participants in the management process. The collective oral history presented here was recorded in 1989 at the Johnson Space Center's Gilruth Recreation Center in Houston, Texas. It includes the recollections of key participants in Apollo's administration, addressing issues such as communication between field centers, the prioritization of technological goals, and the delegation of responsibility. The following people participated:

Howard W. (Bill) Tindall Jr. was responsible for planning all 10 Gemini missions. He was an expert in orbital mechanics and a key figure in the development of rendezvous techniques for Gemini and lunar trajectory techniques for Apollo. He was also the inventor of "Tindallgrams," memos that captured the details of Apollo operations planning. He retired from NASA in 1979.

George E. Mueller was NASA's associate administrator for manned space flight from 1963 to 1969. As such, he was responsible for overseeing the completion of Project Apollo and beginning the development of the Space Shuttle. He left NASA in 1969.

Owen W. Morris worked at the Langley Research Center from 1948 until the Space Task Group moved to Houston, Texas, in 1962. He worked for NASA during Apollo's entirety. Morris was chief engineer of the lunar module, manager of the lunar module, and later the manager of the Apollo program office.

Maxime A. Faget joined the Space Task Group in NASA in 1958. He became NASA Manned Spacecraft Center's (designated the Johnson Space Center in 1973) assistant director for engineering and development in 1962 and later its director. Faget contributed many of the original design concepts for Project Mercury's spacecraft and played a major role in designing virtually every U.S. crewed spacecraft since then, including the Space Shuttle.

Robert R. Gilruth served as assistant director at Langley from 1952 to 1959 and as assistant director (manned satellites) and head of Project

Mercury from 1959 to 1961. In early 1961 an independent Space Task Group was established under Gilruth at Langley to supervise the Mercury program. This group moved to the Manned Spacecraft Center, Houston, Texas, in 1962. Gilruth was then director of the Houston operation from 1962 to 1972.

Christopher C. Kraft Jr. was a long-standing official with NASA throughout the Apollo program. In 1958, while at the Langley Research Center, he became a member of the Space Task Group developing Project Mercury, and he later moved with the Group to Houston in 1962. He was flight director for all of the Mercury and many of the Gemini missions and directed the design of Mission Control at the Manned Spacecraft Center (MSC), later designated Johnson Space Center. He was named the MSC deputy director in 1970 and director two years later, a position he held until his retirement in 1982.

The valuable perspectives of these individuals deepen and expand our understanding of this important historical event.

This gathering was organized through the efforts of the Lyndon Baines Johnson Space Center in Houston, Texas, at the time of the 20th anniversary of the Apollo 11 landing. In particular, Joseph P. Loftus, Jr., played a central role in bringing these key Apollo managers together.

This is the 14th in a series of special studies prepared by the NASA History Office. The *Monographs in Aerospace History* series is designed to provide a wide variety of investigations relative to the history of aeronautics and space. These publications are intended to be tightly focused in terms of subject, relatively short in length, and reproduced in an inexpensive format to allow timely and broad dissemination to researchers in aerospace history. Suggestions for additional publications in the *Monographs in Aerospace History* series are welcome.

Roger D. Launius
Chief Historian
National Aeronautics and
Space Administration
April 18, 1999

Table of Contents

Preface and Acknowledgments

The idea for getting on the record the recollections of those who had been intimately involved in the management of Project Apollo came from Joseph P. Loftus, Jr., of the Johnson Space Center. Mr. Loftus, who had himself been involved in the Apollo project, has long been an advocate of the value of recording the history of space exploration. He, with other Johnson personnel, organized this set of recollections with key personnel from the Apollo management team. The purpose was to exact lessons learned in management practices. I was honored to be asked by Mr. Loftus to moderate this fascinating discussion.

These reminiscences took the form of a workshop that was recorded on videotape at the Gilruth Recreation Center at the Johnson Space Center, on July 21, 1989. This document and the videotape of the workshop itself were made available to various archives and research centers concerned with space and with the major events at that time.

The workshop would not have been possible without the financial support of the Lyndon Baines Johnson Space Center. The ability of the Space Policy Institute to undertake worthwhile projects such as this is a result of the generosity of the several corporate contributors to the Institute's work.

Of course, all of us involved in organizing this workshop owe great thanks to the participants, both for sharing their experiences with us and for the contributions they have made to their country, both during Project Apollo and throughout their careers.

John M. Logsdon
Director
Space Policy Institute
George Washington University

Roundtable Discussion

DR. LOGSDON:
My name is John Logsdon. I am the director of the Space Policy Institute at George Washington University in Washington, D.C. and I have a unique opportunity this afternoon to moderate a discussion among the people that made the Apollo program happen.

Our goal this afternoon is to get down on video tape for both the current and future generations involved in the space program some sense of the working of the Apollo program: how it was managed, how the presidential goal of landing man safely on the Moon and returning him to Earth before the decade is out was turned into an operating program—engineering, development, and operations.

At my left are six of the people most totally involved in the Apollo program. And to get started, I am going to ask each one of them to identify themselves, how they came to the program, and the roles they played in it.

MR. TINDALL:
My name is Bill Tindall. I started with NACA way back in 1948. I got involved at Langley Research Center with Project Echo, and then Mercury, Gemini, and Apollo. I worked for Chris Kraft, who you will meet when we finally get around to the other end. I was a deputy division chief in his Flight Operations Directorate, but the division chief didn't really need a deputy, so most of the time Chris just loaned me out to other places. I worked for Joe Shea for a while on the Apollo on-board software, for the spacecraft software, and then for George Low doing mission techniques, which is basically trying to figure out how we were going to fly the mission.

DR. MUELLER:
I am George Mueller. I am president of the International Academy of Astronautics. I was associate administrator for Manned Spaceflight during the Apollo and the Gemini programs. And my role was, I guess, trying to make everything happen at once and only a little faster than people thought was possible. And over time, we managed to do what people thought was impossible. And that was a very worthwhile thing to accomplish.

MR. MORRIS:
I am Owen Morris. Like Bill, I joined the Langley Research Center in 1948, was in supersonic aerodynamic research until the time the Space Task Group moved to Houston. I worked in the Apollo program office all the way through Apollo. I was primarily chief engineer of the lunar module, manager of the lunar module, and then later in the program, manager of the Apollo program office.

DR. FAGET:
I am Max Faget. I also started at the Langley Research Center in 1946 working for Bob Gilruth. The Mercury program was started there under the Space Task Group, again under Dr. Gilruth. And I came over here to Houston with the Space Task Group to do the lunar mission and I was the head of engineering here at Johnson Space Center.

DR. GILRUTH:
My name is Bob Gilruth. I have worked for the U.S. Government ever since I graduated from college as an aeronautical engineer. I started out with airplanes, went through the developments of World War II, and then when space came along, I was in the right place to work on

flying men in space. We started with orbiting people, and then finally we got to going to the Moon. We flew six missions to the Moon. We brought everybody back, and then shortly after that I did other work and I was no longer in charge. But many of these people here kept it going.

DR. KRAFT:

My name is Chris Kraft. I started at age 20 at the NACA. I came there from college, I suppose, in 1945. I worked for Bob Gilruth in Flight Operations testing airplanes, learning what flying qualities were that Bob Gilruth had invented. From that point on, I did a lot of things that I had a lot of fun with until the space program started. Bob Gilruth asked me to join the Space Task Group in September of 1958. From that time on, I was involved with Flight Operations in all of the manned space flight programs. For Apollo, I was the director of Flight Operations and fortunate enough to be intimately involved in the planning of that great and fantastic voyage. And then from '72 to '82, I was the director of the Johnson Space Center.

DR. LOGSDON:

Clearly a distinguished panel. Gentlemen, what I am going to do is direct a question at one or the other of you as we go along, but I think all of you should feel free to chime in. And since we have so much to cover in a couple of hours, let's try to keep our answers not to the normal garrulous length of telling stories, but really down to the point.

I am going to start with Dr. Gilruth. When John Kennedy went before Congress on May 25, 1961, and said we were going to the Moon, our total flight experience was one 15-minute suborbital flight. You have been widely quoted as being aghast at the notion that the United States was committing itself to send people to the Moon, at least on a specific schedule. Talk a little bit about the kind of challenges that you all of a sudden had responsibility for carrying out.

DR. GILRUTH:

Well, let me tell you this first. The president talked to me before he made his statement. I told him that I thought that maybe we could go to the Moon, but I wasn't sure that we could. And there were a lot of unknowns that we would have to uncover before we were sure. And he said well, let's go ahead and say we can do it in a decade. And we will do the best we can, and if all things work, why we will do what we want to do.

He was very impressed with the reaction of the public to the flights that we had made. And he wanted to do something even much greater than that. Of course, he was a young man. He was much younger than I was at that time. But he was very bright, and he was an easy, good man to work for because he really wanted us to be successful. So that's really how it all got started.

DR. LOGSDON:

Max, I know you had been working with groups in Headquarters and involved in some committees, even before the Kennedy announcement, thinking about the lunar landing as the appropriate post-Mercury goal. What was the kind of engineering outlook at that point? Did we know how to do the job?

DR. FAGET:

We knew what had to be done. How to do it in 10 years was never addressed before the announcement was made. But quite simply, we considered a program of a number of phases.

The first phase was simply to fly out to the Moon, make a circumlunar flight, as we put it, never going into orbit but passing nearby and then whipping on back to the Earth, which was pretty easy on the total propulsion requirements, and was fairly safe from the standpoint of guidance. If you missed the Moon by a large enough margin, you were fairly certain to come back in at an acceptable entry angle.

After doing that, we finally learned a little bit more about what deep space flight might be like. We thought the next phase would be to orbit the Moon, and we would do that with the sense of a follow-on program, that they evolved from the first one and that finally we would evolve to when we would accomplish the landing from a lunar-orbiting, know-how base.

DR. LOGSDON:
But all of that got very compressed very quickly.

DR. FAGET:
It all got compressed into one big program. But actually, in concept, we continued to maintain the idea that we would orbit the Moon before we landed, and we would want to do some reconnaissance from orbit before we landed. Fortunately, there was enough money to buy a lunar orbiter, which was an unmanned spacecraft which provided some excellent photographs of the Moon from which a lot of missions could be planned.

DR. LOGSDON:
All of a sudden, instead of being a group of people that all knew one another located in Hampton, Virginia, you were sitting on top of a national priority, with not exactly a blank check, but with certainly a lot of resources available to you, a set of dates to meet, and instead of doing the work yourself, managing a lot of contracts. I think Bill or Owen may be the right persons to start on this, but I think it goes to all of you. How did you change the character of the work you were doing in order to take on a task of this size? What were the early steps? Owen?

MR. MORRIS:
Well, it was a big change, I think, for just about everybody involved in the program at that time. First, we had to work with Max and his people to understand in a little bit more detail what the systems would be. And we spent almost two years, as a matter of fact, before the final mode

to reach the Moon was selected. And the spacecraft components could then be defined in some detail. In the meantime, we knew enough to start the Command and Service Module since it ultimately, essentially, had to go to the Moon, take care of the people on the way there and the way back, and reenter the Earth's atmosphere. We were not intimately involved with the rest of the operation.

Once we had defined the mission and worked with Chris and Bill and their people to understand a little bit more about the operations, we were able to organize the system a little bit better. I think the biggest challenges that we had, or at least that I, from where I saw the program, was one of communication and coordination. I certainly had been used to working on the smaller jobs, smaller programs, where you intimately knew almost all of the people involved and were able to, by personal relations, do most of the management things you needed to do. All of a sudden, we were thrust into a great big program with tens of thousands of people involved. And trying to get communication and organization set up so that everybody understood how the program worked was probably the biggest challenge.

DR. LOGSDON:
Bill, do you have anything to add to that?

MR. TINDALL:
Well, yes. I think the other thing that you have to remember is that when we really started the space program, it was before Apollo. And we were working on Mercury, and I worked on Echo. It was at that time where people like me, mechanical engineers who didn't know anything about that sort of thing, suddenly found out what orbital mechanics were, how computers worked, how to program them, and things like that. So that it wasn't just a jump right into Apollo, but we really had some pretty nice stepping stones to kind of learn our way along.

DR. KRAFT:

I think that is a good point. I think that you have to recognize that those of us who came from NACA weren't totally blind to the industry. All of us that worked in NACA were used to working with the aircraft manufacturers, particularly we in Flight Operations, Max in missiles, and Bill in instrumentation, and so forth. So it was not a dumb thing to us. And as Bill points out, I think that in the beginning of the Mercury program, where we had guys like Zimmerman to help us with formulating contracts and getting those contracts set up, etc., we weren't totally ignorant to how to do that job.

I think the other aspect of Apollo was that it gave all of us young guys, like Max and myself, under the tutelage of Bob Gilruth, the new opportunity to go out and become managers of this sort. It was a great challenge to us, and we looked at it that way. And it became a heck of a great game for us.

DR. LOGSDON:

So it really started with Mercury.

DR. KRAFT:

Absolutely.

MR. TINDALL:

Even before that.

DR. LOGSDON:

Where you had a set of experiences that you could apply to a much more sizable problem.

DR. KRAFT:

Yes. And I think we learned a great deal about how to approach the spacecraft operations and manufacturing job from Mercury. The industry grew just as we did. So we both grew up together.

DR. GILRUTH:

You must remember that we started Mercury before there was any thought of going to the Moon. And we flew Mercury before that. And it was Gemini, the second spacecraft, well in hand when we realized that we were going to have to go to the Moon. We used Gemini as a way of finding out whether we could do rendezvous and all those other things. So we were kind of lucky that we got started the way we did.

DR. KRAFT:

Very fortunate. I think that the question is did we have to manage a new set of ideas and where were we going in the industry. I don't think these are fair questions because we had just as much a leg up on how to manage the industry as the industry did on how to respond to us.

DR. LOGSDON:

Indeed. Well, George, you got to Headquarters in 1963 and you found this group of strong-willed people by now in Houston, probably pretty well convinced they knew what they were doing.

DR. MUELLER:

They were just starting in Houston.

DR. LOGSDON:

And your job was to make them part of an integrated whole, together with Marshall and all of the other centers, including Kennedy, that were involved with the program. Do you want to reflect on what you found when you got there?

DR. MUELLER:

Well, let me start earlier, because we were all talking about how we got to where we were. And I actually started in the space business with William Wooldridge, back when ballistic missiles were just starting to come into being. And so my own involvement with the space activities was to be in charge of building the first of the lunar probes, Pioneer I, which unfortunately never made it to the Moon, but at least it taught us a good deal about what one

had to do in order to accomplish something like that.

So when I came to NASA, I had a fair background on the commercial side of the field, as it were, trying to sell NASA some of these marvelous new devices that they really needed in order to carry out the lunar mission. And perhaps my major contact at that time was not with the Manned Spacecraft Center, but with Marshall Space Flight Center and working with Wernher von Braun and trying to convince them that they needed to have something that we called systems engineering, or an understanding of the total system and the interfaces between the launch vehicle and the spacecraft and the launch complex, and what needed to be done in order to make sure that those interfaces, when they came together, met the needs of the overall mission. And that, of course, was a key in the long run to the success of Apollo because we did set up a very deep and strict interface control system that made sure that when you delivered things to the Cape, almost

always they fit together. And that, I think, was the most important thing.

So I brought to the program a background in quite a different arena, that was the Air Force management arena. And that combination of the NACA strengths, the old Marshall strengths, and the Air Force experience, I think, really led to the ability to complete the mission within the time scale that we had. And I don't think there was any doubt that we could have carried out the mission, given enough time. But to do it within the decade was more of a challenge. In the long run, it also permitted us to build a team that I don't think has been equaled in the world before or since.

DR. LOGSDON:
But I think it is fair to say that when people here and in Marshall were first confronted with your approach to things, like all-up testing and management of the systems level, there was an initial skepticism that that was the right way to do business.

Dr. Robert R. Gilruth, director, NASA Manned Spacecraft Center, second from left, in Antarctica at "Project Deep Freeze," with (left to right) Dr. Faget, Dr. von Braun, two Deep Freeze scientists, and Dr. Ernst Stuhlinger. (NASA Photo 77-12818.)

DR. MUELLER:

It was a little more than skepticism, I would say, but downright disbelief. And, of course, there were various ways of managing the program. One of the things that I remember very distinctly is the idea of inserting a program management structure in parallel with the functional structure of the centers. And the amount of time it took to convince people that that was, in fact, a good thing to do, and, in my view at least, was necessary in order to provide the kinds of communications that were required in that complex a program in order to be sure that all those interfaces worked.

And in fact, in order to ensure that communications structure, for one thing, and for another to be sure that people were responsible for the important functions that were within a program, I created this idea of five boxes, the five-box management structure, which I don't think was ever widely appreciated, but the idea was to focus, early on in the program, on the fact that you were going to test things, and you ought to design so you can test them. And you are going to have to have reliability, so you have to design for reliability.

And you had to have a system, so you had to set up the interfaces within that system clearly and fix them so that everyone understood what those interfaces were. And you had to have program control, so somebody was keeping track of scheduling dollars and what the implications were. And finally, you had to have someone who was worried about when you had all of this put together, if it will fly and how you will fly it, so we had to have an operations box. And we duplicated this down through the structure in such a way that there were communications between like disciplines so that you could be sure that there was the right set of information flowing up and down the chain in order to be able to make decisions and to follow the program and be sure that everybody was in sync. I think it worked very well.

DR. LOGSDON:

Dr. Gilruth, in setting up the organization to do the Manned Spacecraft Center part of the job, what did you have to do? How were your priorities set? How did you discipline this team to focus on the job?

DR. GILRUTH:

Well, first we had very good people. And I think our task was pretty straightforward. We didn't have to do the big rockets. That was done by another center. We would have a meeting every month where we would tell each other what we were doing and what our problems were, and made sure that we were going to have good interfaces.

It wasn't just by luck. We did all of these important things, and we were lucky enough that if we left one out, or so, we found it before it was too late. And I think we had a very good system. And the big rockets, we couldn't have had a better bunch of people to build those big rockets than we had there.

DR. LOGSDON:

But the relationships between Houston and Huntsville were not always amicable.

DR. GILRUTH:

Yes, they were. They were quite amicable for big organizations like that. We had good friends at the top, and the people got together once a month and we, each, told what our problems were and what we were worrying about in our own things and what we thought we were worrying about in their own business. So we got this thing all worked out pretty well every month. And I generally sat in on those myself. So I think we had a good set of works and a good feeling, I don't think we had any of that where you sometimes see where you have two big groups working together, supposedly. We were good friends with the people there and they were good friends with us.

DR. LOGSDON:

The five of you, other than George Mueller, came from a NACA background. But clearly, those weren't enough people to do the job. You had to do a personnel buildup real quick. Where did the people come from? What were you looking for in your various units? What kinds of skills were there? Was there a kind of personnel base sitting, ready for this kind of task?

DR. GILRUTH:

Well, I remember they shut down a lot of things in England and we were able to get a good number of aeronautical engineers with English backgrounds but were very bright.

DR. MUELLER:

From Canada, I think. Yes.

DR. GILRUTH:

Very bright. And they didn't have to unlearn a lot of things that they might otherwise have had to do. And we got enough people. It was such an exciting program that a lot of people wanted to work on it. Our problem was to make sure that we took the best people and I think our people did a good job at that.

DR. LOGSDON:

Anybody else want to comment on that?

DR. KRAFT:

Yes, I would like to comment on that. I think that in terms of manufacturing and designing and developing, we tried to hire engineers that had 5 to 10 years of experience. And we hired a lot of people. Max hired a great number of people. We had the foresight to get people around him to do our kind of work. And along with the people that we had from Canada, I think that was a good nucleus, the organization.

However, in Operations I looked for people right out of college. That is where I wanted them from. We, frankly, didn't know quite what we were going to have to do and what we were going to have to learn to do that job. And I want you to know that the average age of my organization in 1969 was 26. So, we couldn't have gotten very many guys with a lot of experience.

We had a lot of guys that had some good experience at the top. I, at that time, was about 37 or 38 years old, and Bill Tindall, and Sig Sjoberg about the same. So we were all within the nucleus of the NACA, and the rest of the guys we got out of college.

MR. TINDALL:

True. The exact answer was we got them from Oklahoma and Texas. Everyone wanted to come down here. That's the truth. And they might have had one year of experience at China Lake or something like that. And they all poured in. But none of us knew anything about how to really do the job.

DR. LOGSDON:

You all learned together.

MR. TINDALL:

Yes. And there weren't any courses on that in college on this stuff. There were chemists—

DR. KRAFT:

The other thing I want to stress is we weren't ignorant to organization in NACA. The aircraft industry was run by a matrix organization. We interfaced with those organizations. We knew how they operated. They had functional organizations, they had operations organizations, and they had flight test organizations. We were very familiar with that. And Bob had grown up with that sort of thing in his experience. So we had good knowledge of how to organize, in a functional sense, to get the job done both in research and design and development and in operations. So we had, as an example, Bill Tindall, particularly, and myself, in a management sense had to build a world network. And there was a lot of money for a big system to send things around the world, and contracts around the world to build something that we didn't set out to know about, in the beginning. But we designed that with Western Electric in the Bell labs. So that was a good management experience also.

MR. TINDALL:
Can I say one thing?

DR. LOGSDON:
Sure.

MR. TINDALL:
Another thing that I think was extraordinary, and this was throughout the whole manned-space flight program, was how things were delegated down. I mean, NASA responsibilities were delegated to the people and they, who didn't know how to do these things, were expected to go find out how to do it and do it. And that is what they did. It was just so much fun to watch these young people take on these terribly challenging jobs and do them.

DR. KRAFT:
And that stemmed from the top. Bob Gilruth was that kind of a manager and he taught Chris Kraft and Max Faget how to manage in the same sense. We ended up giving a guy a job and giving him the responsibility to go do it.

DR. FAGET:
Right.

DR. KRAFT:
And I think that everybody had that feeling. And the other thing you have to realize is there was a tremendous feeling of openness among our organizations. We grew up telling each other we were making mistakes when we made them. And that is how we learned. It was extremely important for us to say the mistakes we made as we made them because that helped us to grow. And that feeling was very much a part of our organization.

DR. LOGSDON:
How did that look from Headquarters, George? Did you feel that you had a group that was learning as it was growing and building confidence in being able to do the job?

DR. MUELLER:
Oh, I think we had a group that was not only learning, but was very, very capable. And truthfully, that was the secret of success in NASA, the capability of the individuals involved in all of the centers because we had some tremendous people down at the Cape as well as in Huntsville. I think that one other thing that was instrumental was the fact that we were able to work quite openly with our contractors, a situation which is not true today.

DR. LOGSDON:
Very much so.

DR. MUELLER:
It was so important to be able to actually work with them and to share their problems and be sure that we knew it well enough so that not only did they share their problems, but they felt confident about sharing their problems. Without that confidence you have got a very real problem in terms of getting a program done in a reasonably orderly fashion.

DR. LOGSDON:
I want to talk to the program people about that, but let me push you one more step on that. How did you create a political climate in Washington that allowed that relationship to work?

DR. MUELLER:
You know, it was interesting. We had in Congress some very strong support, and we had set up some guiding committees that provided an oversight. And of course, today you couldn't set up an Apollo execs group and probably couldn't even set up the science and technology advisory group under Charlie Townes because, after all, there would be

conflicts of interest you wouldn't believe in doing that.

And yet, the Apollo execs were very powerful, not so much that they did everything themselves, but they got their organizations geared around to supporting the activity. And the key to that was a real understanding on the part of Congress of what it was we were doing. Every month I met with Tiger Teague and his committee to tell them where we stood, what our problems were the last month, how we were working around them, and whom we were working with. So he was well aware of the Apollo execs group and all of those other things.

At that time, you didn't have the same kind of constraints in dealing with contractors that we do today. I don't think there was a single instance of a contractor taking advantage of this relationship. In fact, if anything, we took advantage of that relationship to get things done that we otherwise could never have gotten done.

DR. KRAFT:
And I think that is one of the most broad points that we have made since we started talking. In our career as NACAers, and the then first years of the NASAers, we encouraged a transfer of people between the industry and NACA. And if you looked across the industry, say take 1960, you would find an awful lot of NACA people in top level management positions in the industry and, likewise, a lot of people from the industry that had come in to NASA to help us manage these programs. That was encouraged. It ought to be encouraged today because it was one of the fundamental strengths of NASA in those early days.

DR. MUELLER:
Exactly.

DR. LOGSDON:
Owen, what about trying to run a contract to get a spacecraft built? What was it like in that environment?

MR. MORRIS:
Well, it was pretty hectic at the time. There were the same kinds of meetings that everybody is talking about, where you would get together on a frequent basis comparing notes about what the problems were, what anybody could do to help the situation. And there was a spirit of cooperation pretty much throughout the program.

There was a little bit of rough edges initially when the contracts were first let, but those were rubbed off quite early. From then on, it was really a team operation. And I think the point I was making earlier about communications, I think, was one of the bigger issues throughout the whole program. Being able to build a team and get people to talk with each other, and get a response out of government, or industry, or the services, or whomever could give a hand, were some of the keys in being able to do it in the kind of time span that we had.

There were a lot of program management techniques learned that were foreign to many of us who were primarily from engineering and research backgrounds. I think those techniques were developed in the early '60s, and by '63 or '64 they were pretty well-honed and in place. And the organization that Dr. Mueller talked about, the five boxes throughout the NASA structure and in most of the contractor structure, also gave a very good point of contact, kind of an input. Anywhere you wanted to go within the organization there was a counterpart whether you knew him or not. Whether you had ever met the man, you knew that if you called that box, he had that same kind of responsibility and you could talk to him and get communication going.

DR. MUELLER:
You know, one thing I would like to remark on is contract structure. One of the things we did was convert the contracts we had, some of

them were cost plus and some were fixed price, into incentive contracts. Incentive contracts perhaps are not used today as we used them. We used them as a means of communication, hard communications. Because we set what we wanted to incentivize, that got the attention of our contractor structure. And it also got our attention because we had to think through exactly what it was we wanted.

MR. MORRIS:
Well said.

DR. LOGSDON:
Max, how did your engineering organization here relate to the technical skills of the contractors? What was the balance of design and engineering choice?

DR. FAGET:
Well, in both the Mercury and of course the Gemini and Apollo programs, I think we were ahead of the contractors. As a matter of fact, before we even put the RFP out, we pretty much knew what we wanted and stated it. The Apollo Command Module was designed more or less by our people.

DR. LOGSDON:
No elaborate Phase A, Phase B kind of structure?

DR. FAGET:
Well, we had a contract with industry to look at the Apollo design and we ran an in-house design at the same time. We had three contractors doing what would now be called a Phase B contract. We ran our own in-house design. And, of course, we kind of took advantage of what our contractors were doing during this time and were continually taking the best parts of their designs and putting them in. So when the final design came out, it didn't look like any of theirs, but was one we had confidence in.

We made some very fundamental decisions during that period of time. One of biggest fundamental decisions that very few people appreciated is that we decided, firmly decided, that we would not use pumps to operate our rocket engines and that we would use hypergolic propellants, just simply because those had all of the characteristics to provide the greatest amount of reliability.

We got criticized roundly by a number of contractors and others, particularly the engine people. They had designed all these wonderful pump-fed engines. Pratt & Whitney in particular had a beautiful engine that ran on hydrogen and oxygen. And the engine demonstrated reliability. I remember a number of times they came to fuss at me and asked why aren't you using my engine. And I said the reason I'm not using your engine is that the engine is not the propulsion system. The propulsion system includes the hydrogen and the oxygen and all of that, and I don't know how they keep all that stuff ready to be used throughout a mission that has got a number of firings. The engine will work, but I don't know that the rest of the propulsion system will work. So we deliberately used less than the highest performance because it provided the simplest in a system where we could have the greatest reliability. We made a number of decisions.

In other areas, NASA was well-funded then. Our development laboratories were well-funded. We had what we called back-up systems being developed in our laboratories, directly under contract from the engineering department. And quite often we ended up having the contractor go to another subcontractor. They had the back-up system, instead of going with the original simply because the development wasn't going along as well. So the engineering organization in many ways kind of stayed ahead of the technology, or at least even with the technology, and was ready to stand in there and plug up some of the deficiencies that occurred during the development program.

DR. KRAFT:

I think one of the biggest strengths of the Johnson Space Center was what Max was just referring to. I would like to have him speak a little more to the fact that what we were doing with the finances we had was building ourselves a team of guys that were just as good or better because we were hands-on people. I would like to have Max talk to that because it is an extremely important point.

DR. LOGSDON:

That was the point I was driving at. Whether you got pressure to change from the outside, or whether the contractors said hey, these folks are telling us how to do our business, that strong engineering core was essential to the ability to carry off the mission.

DR. FAGET:

Yes, we did get pressure to change it. And I am going to be very frank. I think most of the pressure came from Headquarters. Headquarters wanted to do studies, we wanted to do development. So we spent an awful lot of money on studies, and we studied things to death --

DR. LOGSDON:

With not much intent to do anything about it.

DR. FAGET:

With the Apollo program. I can show you studies of space stations, and everything else that was done back then. And it didn't lead to anything because there wasn't enough money to follow through. And when we did want to do something, quite often the most practical system to do it was nonexistent because the development work wasn't done. The NACA did an awful lot of development work in that laboratory. Of course, I came from that background. I am not going to say that I am completely right, but I know one thing in my own mind. In my own mind, that is the best way to do it. And we didn't do it that way after a while.

DR. KRAFT:

Max, let's go a little further there. I think that what we recognize, from Mercury to Gemini to Apollo, was that we needed a certain percentage of the funds to do these kinds of things in Max's laboratory which gave our people straight, hands-on knowledge, first-hand knowledge, and it allowed us to build systems that we built at the time of Mercury to use in Gemini and at the time of Gemini to use in Apollo. The perfect example of that is the fuel cell. The fuel cells that we used in Apollo were developed with Gemini money.

And so it was extremely important. The fuel cells that we used in the shuttle were developed with Apollo money. And that was a concept we all preserved, all of us sitting right here said it was a great thing to do because it gives our guys the great knowledge of how to build these systems and work on designs. So they are just as good as the guys out there that are doing it at the same time. That was a tremendous thing. And we just weren't able to continue and that really hurt NASA.

DR. LOGSDON:

Why not? Is it basically a question of money or is there something that underpins the money?

DR. FAGET:

It is a question of money because he who has the money makes the decisions on how it is going to be spent. The money in the present NASA system is spread out among the organizations at Headquarters, they in turn have their channels of spending the money, and ultimately a lot of the say of how that money is spent is made by people up at Headquarters without much hands-on experience. They have a lot of theoretical experience. They are well-educated. I know that. But very few of them have ever really done anything and they are making all the decisions on how the money is being spent.

Left to right–Eugene F. Kranz, deputy director of flight operations; Christopher C. Kraft; and Maxime A. Faget at Mission Control in Houston. (NASA Photo 81-H-365; S-81-30139.)

DR. KRAFT:

We can't over-emphasize that point because that was the strength of Marshall, of KSC, and ourselves that they worked in that kind of a fashion. And it dried up post-Apollo. And it really hurt us, and it was because of money because we didn't have enough to support the program itself. So when we started trying to siphon off some of that for future technology, we could not do it. It just went away.

MR. TINDALL:

We didn't hire anyone during the entire '70s, for crying out loud.

MR. MORRIS:

Might be made along these lines. One was that they were saying a lot of the hardware built for Gemini came from Mercury development, Apollo built on Gemini, the Shuttle built on Apollo. The financial support was much bigger in those days, and the guy managing the Gemini program could afford to invest in developing better systems which he might use or the next guy might use.

As the budgets got tight in the late '60s and on into the '70s, the program manager was faced with the problem that he either had to put the money directly into his program or have his program suffer materially to foster development for the next program coming down the line, and it was very easy to see where he was going make his decision. I think that is one of the problems the agency has had.

DR. KRAFT:

You probably have a lot to say about that because I think you had a lot to do with making sure that took place. I mean, you understood that was a necessary element of what we were doing and you could see that as good expenditures of funds.

DR. MUELLER:

I have been surprised that there has been no recognition in NASA over the years that the important thing is the investment in technology, because that is where we, looking back now, I had the opportunity of going around and seeing what has been done in the last 20 years. And in the Apollo era, one of the things that we did insist on was sufficient funds, free funds from the centers, to make decisions about where to put development money. And so we had, essentially in every center, an ability to spend money on development of future things. Looking back, going back 20 years, what I find is that over that period of time, somehow or another, all of our engineers have become contract managers and almost none of them now go out in the lab and do some work.

MR. MORRIS:
Amen. Yes.

DR. MUELLER:
And even in our so-called research centers, they have become contract managers.

MR. MORRIS:
Amen.

DR. MUELLER:
And they keep going out and getting contracts instead of going out and doing technology. And that is a serious weakness. I mean, it affects the whole structure of NASA in a way that, I think, is very detrimental.

DR. KRAFT:
Very profound. Very profound.

DR. MUELLER:
Yes.

DR. LOGSDON:
Let's talk about risk management for a little while, risks both in developing the systems and then operating them. What kind of attitude permeated the organization as you approached this task? How did you make your trade-offs? I don't even know who to ask to start that discussion. I am sure you were all involved.

DR. FAGET:
There was a great deal of risk.

DR. GILRUTH:
In what we were doing. No question about it. Something brand new in a place man had never seen, man had never been. And it was all new and it was tough.

DR. MUELLER:
And we had a lot of help, those who helped us to do reliability analysis and prove that everything was going to work perfectly.

DR. GILRUTH:
Yes.

DR. MUELLER:
Or else we wouldn't fly.

MR. TINDALL:
But we also had backup systems for everything, or almost everything that we could. And I know when it came to our planning missions and procedures and all of the software in the mission control center and things like that, at least, it must have been 80 percent or maybe 90 percent was spent on nonnominal situations. Everyone was trying to figure out what are you going to do if this happens, what are you going to do if that happens, work through the system rules, if the system was working or not working, and how do you decide, and all of that. And not just one source of information, but usually we would triple or quad it.

DR. KRAFT:
You know, the thing that I want to say about this risk business is that I think this numbers game, as George is implying, was greatly over-played, is greatly over-played. And the way I felt about it was the following, and I think Max would back me up. We said to ourselves that we have now done everything we know to do. We feel comfortable with all of the unknowns that we went into this program with. We know there may be some unknown unknowns, but we don't know what else to do to make this thing risk-free, so it is time to go.

DR. GILRUTH:

That's right. That's right.

DR. KRAFT:

And I can't say it any differently than that. I think if any of us, Bob Gilruth from the top, myself, or any of us, Max Faget felt, well we don't know what we are doing here. We don't know the answer to that and we should know the answer, there was no question in our minds we weren't going to do it no matter what. We were going to wait. We'll wait and we'll wait and we'll wait. But when we feel like we are ready to go, to hell with this risk analysis business. We have done everything we can. Let's go do it.

DR. LOGSDON:

Are there examples where you did wait?

DR. KRAFT:

Of course.

DR. FAGET:

Oh, sure.

DR. LOGSDON:

Talk about a couple.

DR. FAGET:

Some of the unknown risks, of course, we dealt with by making unmanned flights to test out the systems, parachutes and things like that. We realized you make a parachute that's 60 feet in diameter and you are going to deploy it at one time, that was something kind of new. And if you are going to hang-glide three people on that thing, you better damn well test it. And so we had a very sensible, thorough test program to make sure that those parachutes would work.

When it came to risk management, I always had an awful lot of trouble with some people coming in and telling me about redundancy and everything else. I always took the attitude that gosh, I am supposed to be an engineer. And if I am an engineer, I better damn well understand what reliability and what failure means, otherwise I am not an engineer. And I expect the engineers that work for me to take the same attitude.

And that goes back to the propulsion systems. We took exactly that attitude before we even had the specifications for RFP, we had decided on what kind of propulsion system we were going to use because we could go through that. And by picking the pressure-fed hypergolic propulsion system, we could see the least number of conflicts. And then we said okay, you can't have redundant tanks. You simply can't carry twice as much propellant, fire what you need and throw away half of it. It is pretty hard to have redundant thrust chambers. But everything else besides the tanks and the thrust chambers and some of the propellant lines were all redundant. We had redundant valves, quad-redundant valves, everything else. Check valves, you bet we had those things redundant. And from that standpoint, the same thing in the pressure feed system. And so I basically said the best way to deal with risk management is in the basic conceptual design, get the damn risk out of it. And I think that is what made the program a success.

DR. MUELLER:

I guess I had been point man on this reliability analysis since the Apollo days because I got involved with the president's science adviser in his insisting that we do a complete reliability analysis. You have to look at it in two ways. One is that each of these flights is a unique event.

So statistical analysis has a limited utility in trying to define what the probability of success of one single event is. And that is one thing. But another thing, that Max touched on, is that the place to use the reliability analysis is in the design process. You can't measure reliability in, you have got to design it in. And once you have got it designed and it is built in, from then on, whatever else you do is just window dressing, as far as I am concerned. Now, I will temper that, though, by saying that one of the important things one needs to do is to recognize when you have a failure and be sure you really understand in depth what that failure is due to and make sure you have fixed it to make sure it doesn't happen again.

MR. MORRIS:
Amen.

DR. KRAFT:
Part of the creed.

DR. FAGET:
I couldn't agree with you more, George. And one of the things that you want to do is make sure that you have enough testing. You are bound to have blind spots. There isn't any engineer that I know of, God or not, that is not going to make some human errors. So you have to have testing to pick up those blind spots. And if you don't do those tests and if you don't believe what you see in those tests, you are in trouble. And I will give you a good example of when we did that was on the solid rocket motor. That was a blind spot. We designed in failure, and I would not hold a design engineer to fault for that failure system, but I would hold those that came after the design.

So you need a good design. You have to come in with a good, basic design. What you might call testing, and a reliability audit, if I might use those words, as opposed to really saying, you know, it is no good until I approve it is right, as a second man coming in. An audit is good because it gives a second party a chance to look at what you have done. But it ought to be done by good engineering heads as opposed to a bunch of mathematicians.

DR. KRAFT:
I know you don't want us to tell war stories, but I have to tell a story in that regard. We arrived at a management council meeting in Washington around noon. And this guy right here has a bunch of papers sitting in front of each of our desks that said could you please give me the reliability number associated with risk, at each one of the phases of the mission, and they were launch, translunar injection, orbital coast, etc., etc. The only thing that you and I want to remember about those numbers is two things. Number one, George Low and Chris Kraft got the identical number in the total listing of what was going to come, and I am not going to tell you what that number was. And I also want to say that George Mueller tore all those goddamn pieces of paper up and never asked us that again.

DR. LOGSDON:
There was a difference in test philosophy. Dr. Mueller brought with him the idea of all-up testing, testing as many systems as possible at one time. What if you had earlier had a large failure in one of the all-up tests? What would that have done to the program?

DR. MUELLER:
We would have found out what failed, and we did, as a matter of fact, have a real failure. I will say we did. We found out what failed and we fixed it.

DR. KRAFT:
And we put a man on it the next time we flew it.

DR. MUELLER:
Exactly.

DR. FAGET:
Hindsight is a wonderful thing. I thought that was a good idea. Even at the time I thought it was a good idea. But I believe that anything that works is bound to be a good idea. But it is a matter of being bold. It was a bold idea. You know, you can do step by step testing, a kind of sure-fly way. There is a risk. If you do an all-up test and you have a major failure, you might not be able to find out what failed. And that is your risk.

DR. MUELLER:
Yes, that's the real risk.

DR. FAGET:
The stand point of the ultimate risks to the crew in the all-up test provides just as much of a safety audit or a confidence as a bunch of separate tests. And in many ways it is better because it has the systems working against each other. But it doesn't undergo this one risk, which thank God we didn't run into, when we ended up with a big failure and we couldn't figure out what gave it.

DR. KRAFT:
That was a significant thing. That was a characteristic of our organization at the time. Our ability to fly the Saturn V after we had had a major malfunction on the SII stage, which we did the previous time before we flew Apollo 8. And that took a lot of guts, a lot of nerve, but I think we knew what we were doing. And we looked at that and said the risk is worth the gain. That would be extremely difficult to do in the environment that exists today.

DR. LOGSDON:
That is what I would like to pull out a little bit. We are risk adverse now. What allowed you to be bold?

DR. FAGET:
What we are all saying is that there is a time to be conservative and a time to be bold. And judgment, good judgment, tells you when to do it. And of course, we had great judgment.

DR. KRAFT:
But you are right. I don't think there was a soul in any level of management in our organizations that I know of that was opposed to flying Apollo 8.

DR. FAGET:
No.

DR. KRAFT:
I cannot think of a single soul that was opposed to flying Apollo—

DR. MUELLER:
The only ones were the media.

DR. KRAFT:
But I am talking of when we got there. I mean, there were a lot of questions about it until we got there. George Mueller says you guys have lost your minds when we first thought that up. But when we got there and we said we were going to go, there was not a single question as to whether we were going.

DR. MUELLER:
We wouldn't have gone if there had been.

MR. MORRIS:
That's right. If there were a question we would not have done it.

DR. KRAFT:
It only took one voice to say we weren't going.

DR. MUELLER:
Exactly. But let me just say I don't think that we were not risk aversive at the time of Apollo. In fact, we spent a great deal of time, energy, and effort being sure that we understood the risks. So it wasn't that we were just boldly marching out where angels feared to tread. We really understood what our system was.

So today, we may be not willing to take any risk, but in that case, you can't fly because there is always going to be risk. And to think that the shuttle is risk-free would be a major mistake. Even the best of solid rocket technology eventually has a failure. And even the best of liquid rocket technology eventually has a failure. So you have got to expect that you are going to have failures in the future.

Our ability to fly the Saturn V after we had had a major malfunction on the SII stage… took a lot of guts, a lot of nerve, but I think we knew what we were doing.

DR. LOGSDON:
Well, you had a traumatic failure. You lost a crew in the Apollo 1 fire. What did that change?

DR. MUELLER:
It changed our test procedure for one thing.

MR. MORRIS:
Well, we tightened up.

DR. LOGSDON:
But did it change the way you could do your business?

MR. MORRIS:
We tightened up.

DR. LOGSDON:
Had you gotten a little loose?

MR. MORRIS:
Yes, yes.

DR. KRAFT:
I think it changed the way we did our business.

MR. MORRIS:
It sure did. I think the configuration management received a lot more attention after the fire. We had reasonably good configuration management before. In the interest of time, there were a lot of changes made and the paperwork caught up with it afterwards. There were a lot of small changes that were really not reviewed in great detail. As a result of the accident, the fire, the procedures were really tightened up. And I think it was all to the good of the program. I think it helped—

DR. LOGSDON:
Kind of a brutal reminder.

MR. MORRIS:
The program matured much more rapidly from that point on because the rigor was in there.

DR. KRAFT:
Well, I think the other point you have to make there is that, let's admit to the fact that we were running our fannies off trying to do Apollo. And it was difficult for us to take the lessons learned in Mercury and Gemini and apply them back into Apollo as we designed it and as we built it because it was tough, just tough to do from a communications point of view, right?

Now, when we had the fire, I think we took a step back in and said okay, what are these lessons that we have learned from Mercury and Gemini. What lessons have we learned from this horrible tragedy. And now let's pump that back and be doubly sure that we are going to do it right the next time. And I think that that fact right there is what allowed us to get Apollo done in the '60s.

DR. MUELLER:
Right. I think one thing you ought to recognize is it wasn't just that we fixed the fire, we fixed everything else we could find that had any possibility of being fixed.

MR. MORRIS:
Right.

DR. LOGSDON:
And you found some other things that needed fixing.

DR. MUELLER:
A whole large number of things.

DR. KRAFT:
One hundred and twenty-five of them.

DR. MUELLER:
Yes, a large number.

MR. TINDALL:
You know, I am not sure what risk management is. If it is only dealing with the spacecraft systems, well that has been discussed. But I think

we are leaving out something that is pretty darn important, and that is how you operate the vehicle and the people and all of the rest of it after that.

I think one of the greatest contributors to minimizing risk was the extraordinary amount of training that was done, high-fidelity simulations that were extraordinary. And there is no question about it, they saved us. I mean really saved us many, many times because I don't think there was a single mission that we didn't have some significant failures. The fact was that people could figure them out because they had been trained and knew how to work with each other. The communications were there, the procedures were there to figure out what to do in real time and get the thing going. And most of the time when those things happened, the outside world didn't even know about it.

DR. LOGSDON:
Bill, when you came to plan how to actually fly these missions, and maybe Chris will want to comment on this also, were you happy with what the engineers had given you? Did you have systems that were ready to go?

MR. TINDALL:
Yes. There were practically no changes that I can recall that we went back and asked for in the spacecraft systems. I am not saying that there weren't any. There were maybe one or two pretty minor ones, minor ones and easily fixed ones. As far as I could tell, the spacecraft were safe.

DR. MUELLER:
I have a different view of that, as a matter of fact, because all through the design of the spacecraft the astronauts and the flight operators were involved in it. So we had a lot of changes in the design process.

MR. TINDALL:
During the design process, yes.

DR. KRAFT:
It was an iterative process there.

DR. MUELLER:
Right.

MR. MORRIS:
It was not a question of the designers and the builders then turning it over to the operations guys and them saying "what is this stuff?" They had been in it from the conceptual design, from the specification writing.

DR. MUELLER:
Sure, right.

MR. MORRIS:
When we sat down to write the specifications, we wanted the operations people right in the middle of it.

DR. KRAFT:
We argued over the instrumentation we wanted. We had a lot of arguments back and forth.

MR. MORRIS:
Sure.

DR. KRAFT:
But by the time it got to Bill Tindall's technique development, we accepted what we got and used that as a set of mission limitations just as you would in the envelope of an airplane or anything else.

MR. MORRIS:
Right, right.

DR. KRAFT:
But before we got to that point, it was very much an iterative process.

DR. LOGSDON:
Max, you wanted to say something?

DR. FAGET:
Well, I just was going to say in a different way than everybody else that we did have an integrated team. We did have a continuous sequence of design reviews where the engineering people from my organization, the program management team, and the operation team would all participate. And it was really an open forum. The program management team was the only one who would make the decision, but everyone could speak as long as he was unsatisfied with what was going on.

DR. MUELLER:
It's true.

DR. FAGET:
And stated his views. And then everyone would discuss those views, whether they were reasonable or unreasonable. And then the program managers would make a decision.

MR. MORRIS:
That's right.

DR. FAGET:
And that brought a lot of operational consideration into everything that we did from the standpoint of how the systems were designed.

MR. MORRIS:
Certainly.

DR. FAGET:
They were designed to operate, to work, and, of course, to be reliable.

DR. LOGSDON:
Were there any major differences in the command and service module contract and devel-

opment and the lunar module development? They were different spacecraft for specialized functions with different contractors. You hear a lot about the relationship between Houston and North American at the time. You hear very little discussion about the relationship between Houston and Grumman and the lunar module. Is there anything worth talking about there?

MR. MORRIS:
I think probably the biggest reason for that was that the lunar module was started later. The command module had two years of that contract underway before the lunar module was placed under contract. So a lot of the rough edges had been worn off. The team here at Houston and the team at Rockwell was much better at that point in time.

DR. KRAFT:
I don't doubt what you say is true, but what you say is a misconception. I don't think that is right. I think there was just as intimate a feeling between the organizations all across the board at Grumman as there was a—

DR. LOGSDON:
Well, I am not suggesting a detach. It is just that nobody seems to talk about the positive character of that relationship.

Dr. George E. Mueller, associate administrator for the Office of Manned Space Flight, with Lt. Gen. Samuel C. Philips, director of the Apollo Program, in Firing Room 1 of the Launch Control Center at NASA's Kennedy Space Center. (NASA Photo 69-H-1064; 107-KSC-69P-574.)

DR. KRAFT:
Well, let me tell you what we did. We discussed this, because I was at the management council meeting, as a total thing, both command module, service module, software operations. We always did a data dump for George every time we had one of those meetings.

But I wanted to describe another meeting, and that was that once a month, and sometimes more often, George Low had a round robin, and we got on an airplane. He had a Configuration Control Board meeting, and once a month it was held at Grumman and at Rockwell. And both Grumman and Rockwell people attended those meetings. We would fly from here to Grumman, have an all-day meeting, get on an airplane, fly all night, have the same meeting in California with Rockwell and fly home. And that was done all the time. So everybody was familiar with what was going on in all of those contracts as far as we were concerned. A very important thing to happen.

DR. MUELLER:
And Grumman had a fair number of problems also.

MR. MORRIS:
Oh, sure.

DR. MUELLER:
It was just that we worked them a little harder and earlier.

MR. MORRIS:
And we knew how to go about them a little bit better.

DR. MUELLER:
Right.

MR. MORRIS:
We were a little bit more experienced, too.

DR. LOGSDON:
You were learning from what had gone on before.

MR. MORRIS:
Yes, from our previous experience, yes. Sure.

DR. LOGSDON:
Let's talk about flying the missions for a while. You mentioned earlier, a couple of you, the Apollo 8 decision, the decision to commit a crew to a circumlunar, a lunar orbit flight with the first manned mission of the Saturn V. What would have happened if there had been a major problem there. What would that have done to the program?

DR. MUELLER:
Define problem.

DR. KRAFT:
Well, if we had had a similar malfunction on Apollo 8 as we had on Apollo 13, we would have been in a hell of a mess. And God knows what that would have done to the program. Anybody could give his or her own guess to that. So that would have been terrible.

DR. LOGSDON:
How risky did you feel that choice was. You earlier said no one in the organization down here stood up and said no.

DR. KRAFT:
It had a certain amount of risk to it. However, remember that that also was a human failure. There was not a damn thing wrong with the hardware that caused Apollo 13 to fail, it was those idiots that treated the hardware improperly and caused that malfunction to take place.

DR. LOGSDON:
And there was not a thing you could do about that then.

DR. KRAFT:
That's right. As Max just described a little while ago, those are human frailties. But the Apollo 8 thing, I think, was one of the most fantastic examples of good management that I have ever seen.

DR. MUELLER:

I thought that was one of the turning points in the program and one of the best things that we did.

MR. MORRIS:

Sure was.

DR. MUELLER:

Although these guys would never believe it, I was enthusiastic about that mission from the time I first heard about it. But I used it as a lever to make sure that everybody did their homework and we were certain that it was all going to be all right.

DR. LOGSDON:

Were you ready from a flight ops point of view to take on that mission?

MR. TINDALL:

Absolutely.

DR. KRAFT:

Well, we made ourselves ready, I think, is the right way to put it. I mean, when first asked, I probably, in Bob's office, got red in the face. And three days later told him we are "go." So that is the way it went.

DR. FAGET:

I didn't have anything to do with planning that mission, but I can just remember one thing. When that bird went behind the Moon and it was supposed to make a burn in the dark and I had to sit there and wait until it came back into the clear, that was a very exciting, high heart-rate time for me.

MR. MORRIS:

That was a long pass.

DR. KRAFT:

It was particularly for me, too, because the sequence of the thing was as follows. When we got to thinking about doing the thing and

George Low asked us what we could do to go around the Moon, when the ops guys got together, we decided look, we are having a hell of a lot of trouble with this lunar orbit determination. We have been looking at the lunar obiter data, and we can't figure out where the spacecraft is when it comes back around. We are several thousand feet off. And if we are going to be that way for the whole mission, we are going to be in trouble because we won't know where to start the descent.

So we said look, if we are going to do this mission, let's really get something out of it. So let's go and orbit around the Moon. Now, there were a lot of white-faced astronauts when we said that. But nevertheless, we again said that we thought that risk was worth the gain. And it was. That was the flight that we learned an empirical method for doing an orbit determination around the Moon and really put us ahead of the game. Apollo 8 gave us all, I am talking about engineering, program management, top management, operations, it gave us all a tremendous feeling of confidence that we knew what we were doing after that situation.

MR. TINDALL:

We knew what we were doing, we had procedures laid out, we tried to imagine every single conceivable failure that could occur and what we were going to do about them. We did things like lunar orbit insertion in two stages instead of one big burn because an overburn that might go undetected would cause it to crash into the Moon. So we backed off and did a two-burn.

I mean, things like that, that we could tolerate some pretty significant problems. I don't know what would have happened if we had been hit by lightening like we were on Apollo 12, but the fact of the matter was we could have gone ahead and flown. We might not have had the guts later in the program to do that, but Lord.

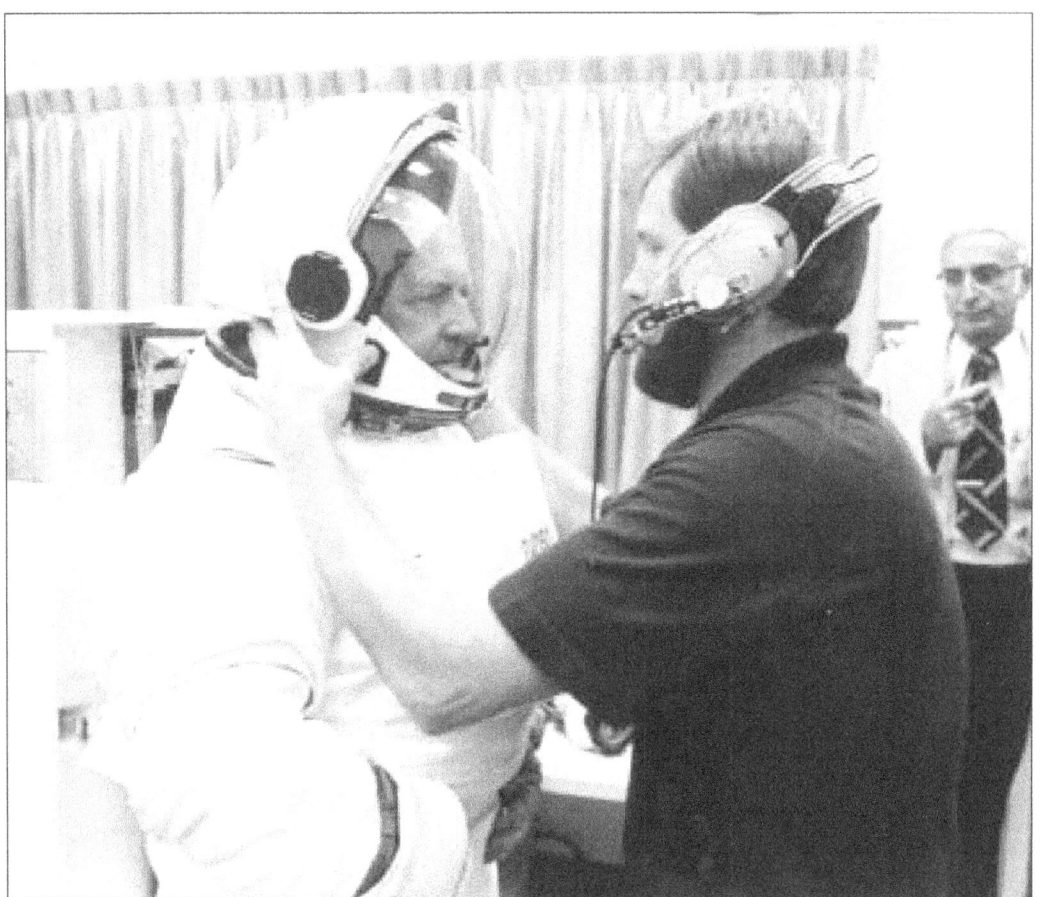

Dr. Christopher C. Kraft, director, Johnson Space Center, July 1976, is assisted into a developmental Space Shuttle pressure garment. The JSC-developed space suit is a two-piece, adjustable to fit, modular design suit, sized small, medium, and large to fit crew members of either sex. (NASA Photo S-76-26409.)

MR. TINDALL:
Kind of like what I was doing. They called me a professional pessimist.

MR. MORRIS:
Yes, right.

DR. KRAFT:
That's another profound thing, in my opinion. I think that the thing we learned, and the thing that made us strong, was that we knew about failure. We recognized failure, we knew it was there, we always looked for it.

DR. MUELLER:
Of course, we had a free return trajectory.

MR. TINDALL:
Free return until you fired the SPS.

DR. LOGSDON:
Dr. Gilruth, by this time were you beginning to be convinced that this was going to work? I say that a little facetiously.

DR. GILRUTH:
I was always a great worrier.

DR. MUELLER:
We all were.

DR. GILRUTH:
I felt that it was much better to be a great worrier than a person that didn't have a trouble. And of course, a person either worries or he doesn't. Some people don't worry about things. I happen to be one that worries very, very much, especially on things like flying men to the Moon. We had a lot of worriers in there and most of them hid it quite well. And I hid it pretty well, too, except when I would be all-alone with some close friends.

MR. MORRIS:
That's right.

DR. KRAFT:
And everything we did was based on decisions on failure rather than success. And if you want my opinion, that is what happened to NASA in the Challenger accident, their decisions were based on success, and the people sitting right here made decisions based on failure. And that may sound crazy as hell, but I believe that is the way we did it.

MR. MORRIS:
That is what we spent our time on.

DR. KRAFT:
That's right.

DR. LOGSDON:
And the attitude clearly is very different today.

DR. KRAFT:
Yes.

DR. LOGSDON:
Why? What were the elements in the climate in which you worked that made it possible to

operate worrying about failure but anticipating success, I guess, and being free to do that, not having to promise success?

DR. KRAFT:
It is what we have been saying. We all, there is not a soul here or was in our organizations that felt like they couldn't say what they wanted to say any time they wanted to say it and felt totally comfortable about it. We grew, that was our heritage. It was the way we thought. We were never embarrassed about it. We were never embarrassed about being made a fool of when we made mistakes because we made them. I mean, we made hundreds of them. But we were used to being open about them. And that was fundamental to getting our job done.

MR. TINDALL:
And in fact, it is exactly the opposite. If someone were to have found hidden problems and not bring them out, that was the worst kind of person to have around.

MR. MORRIS:
Amen.

MR. TINDALL:
Absolutely the worst.

DR. LOGSDON:
And you felt, as a center here, you could take that attitude and Headquarters would let you?

DR. GILRUTH:
Well, we didn't ask them. We did it. We couldn't have operated any other way than that.

DR. MUELLER:
Well, Headquarters couldn't have operated any other way either.

MR. MORRIS:
Well, I think that it rather than injured us, we were encouraged.

DR. GILRUTH:
We didn't have any trouble from Headquarters. They thought we were doing right.

DR. KRAFT:
I don't think that any of us felt the least bit inhibited in saying what we thought. A lot of people didn't like it, but at least they respected us for it.

DR. LOGSDON:
During the program, was money ever a constraint?

MR. MORRIS:
Oh, sure.

MR. TINDALL:
No.

MR. MORRIS:
Not nearly to the same extent as it was on the shuttle and is today on the space station. But yes, money was a constraint, from 1964 on it became more and more of a constraint.

DR. LOGSDON:
What kind? I mean, Bill said no and you say yes.

MR. MORRIS:
Well, comparatively speaking, I would say no. And I would try to qualify that. The number of backup systems that you were able to fund, the amount of hardware that you were able to build, was somewhat constrained. It never got to the point that you violated some of the real premises of the program, and I think that one of the things that allowed us to transcend a number of lines in the reliability industry was the failure analysis that we did. Any time we had a failure, we spent whatever it took to truly understand that failure and truly understand that it was fixed before we let go of it. Those kinds of things we always had money for.

In later programs, instead of testing the hardware, in many cases it was qualified by what was called similarity. It was like another part, an engineer made a judgment that it was close enough or it wasn't close enough, and that was it. That was the end of the testing. There was none. And so they had, later on, those kinds of monetary constraints. And in Apollo, we never did that. But yes, there were monetary constraints that were put on and there was a regime in which you had to live.

MR. TINDALL:
Sure, but it came after the mission success was way at the very top.

MR. MORRIS:
Oh, yes. Yes.

MR. TINDALL:
And schedule was pretty high up there.

MR. MORRIS:
Yes, absolutely.

DR. FAGET:
But there was funding to take.

MR. TINDALL:
Well, we could, I suppose, have used more money.

MR. MORRIS:
Well, we did not do some things that we would like to have done. We were able to do everything that we felt was necessary.

DR. MUELLER:
I can't think of many.

MR. MORRIS:
Well, we got done what was necessary.

DR. KRAFT:
I don't know if I should get within an inch of that. We were the guys that were always bring-

ing in the money problems. And we weren't the least bit inhibited for bringing in the money problems, either.

DR. FAGET:
We always seemed to never have the money for travel and we never had enough billets for the people that we wanted.

DR. MUELLER:
Well, billets were really outside our control. The travel, I have never understood why we ever curtailed travel. But that also was outside of my control at the time. There was some dictum by President Johnson that people shouldn't travel.

MR. TINDALL:
Yes, those damn GSA cars out there in Arlington. I mean, that was so bad.

DR. FAGET:
George, you never heard this, but I remember one time we were having trouble with travel and I was talking to one of your money controllers, I don't know who it was, but I explained to him that we were really in a hard way. And he had some data. He said well, the people at JSC travel more per capita than any other center. And I said well, you have proven another point, we don't have the people.

DR. MUELLER:
True enough, as a matter of fact. Although, the centers' population seems to even out over time, I have noticed. But it is true that people regulating things in Washington often create a regulation that has zero relationship to the real needs of a program, for example. And travel was one of them. We worked our way around that in every way we possibly could.

DR. GILRUTH:
But that wasn't in the way of a budget limitation.

DR. MUELLER:
Exactly.

DR. GILRUTH:
So was the head count.

DR. MUELLER:
But I must say that we always found money enough for some odd things, like that swimming pool that was built down at Huntsville.

DR. LOGSDON:
The neutral buoyancy simulator.

DR. KRAFT:
And we traded monies around in terms of travel, too. We traded monies back and forth among organizations. It was tight.

DR. FAGET:
But we got by.

DR. LOGSDON:
Once you started flying, what were the problems? You had to make a decision, I guess, as to how many missions to fly and what the objectives of each one were. Let's talk about that a little bit.

DR. GILRUTH:
The decision was made before we started flying. And I don't think we had to change it much.

DR. KRAFT:
No, I think George Mueller, in his regimentation of us, set out a class of missions—A, B, C, D— that from the very beginning we all participated in with a great amount of discussion and work. We had certain goals and objectives we were going to get out of each class of missions and set those beforehand. Frankly, I don't think any of us expected that each one of those would be done with one flight, but that's the way it turned out.

We were able to build on our experience and learned from each one of those things and we

didn't have to do that. I think we were very, very fortunate, though, that we didn't. I don't think any of us expected to land on Apollo 11 as simply and as straightforwardly as we did. I think we all expected that we would have to do maybe one or two flights in A and maybe two flights in B and two flights in C. But as we pointed out, Apollo 8 really gave us the turning point in the program to make that differently. That sort of gave us a different mindset, I suppose.

DR. MUELLER:
The only glitch was Apollo 13.

MR. MORRIS:
Right.

DR. MUELLER:
And that was one of those things that should never happen anyhow.

MR. MORRIS:
That's right.

DR. LOGSDON:
But did Apollo 13 hasten the end of the flights?

DR. MUELLER:
Let me just say that the real problem that we had was the support of the scientific community and their feeling that all of this money was being wasted on manned space flight and man landing on the Moon early on. That is before we started flying and landing.

MR. TINDALL:
Yes, very much so. That was a great problem for all of us.

DR. MUELLER:
And then, once we got to landing and were beginning to pick up some data, they became very strong supporters. It was around that time Carl Sagan flipped.

MR. MORRIS:
All of them flipped, I think.

DR. MUELLER:
Yes. But by that time, we had already committed to the end of the program and there wasn't much we could do about it. The Bureau of the Budget had us where we couldn't do anything at that point.

DR. GILRUTH:
We landed six times. That was enough.

DR. KRAFT:
I think that is a point that has been discussed a little bit recently.

DR. LOGSDON:
I would like to push on that point a little bit, that six was enough in his judgment. That perhaps the risk-benefit calculation changed somewhere in there, that the risk of each additional mission might have been greater than the benefits.

DR. GILRUTH:
Well, one of the things we had to find out was whether the Moon was different in different places. And no matter how well you did in one landing, you didn't know what it was like in another place. And we went to three or four different places, and they really were somewhat different, but basic things were not.

We felt that we had gotten about as much as we could get, unless we found something later on. But I think the scientists were pretty well satisfied, too, that we had made a good going of the various places around the Moon and doing the kinds of things as best they could think of what we should do.

DR. KRAFT:
I think that has been discussed in recent weeks relative to our process with the scientists, which I thought George made a great point of.

And that is, on the first mission, we used a lot of profanity if you want to know the facts of it. We want to get the damn job done. Let us fly one time. Let us get familiar with the operation. Let us prove to ourselves that we know what we are doing, how we can get there, and that we have got a familiarity with that. After that, we will do anything you ask us to do within the performance of this vehicle.

And if you look at the record, that is exactly what we did. After Apollo 11, we did not make a single move without the scientists saying that was what they wanted to do. All we did was give them the performance limits, tell them what we thought we could do, and in that envelope. And from then on, they were in control, totally in control.

MR. MORRIS:
As a matter of fact, after the first mission, almost all of our activity was to enhance the scientific capability of the vehicle.

DR. KRAFT:
Absolutely.

MR. MORRIS:
We brought the lunar rover on Apollo 15. We had the extended lunar module. We were able to stay another day. We had more cargo capability.

DR. KRAFT:
We put an automobile on the spacecraft, and they could go 20 miles away from the landing.

MR. MORRIS:
So almost all of our activity after Apollo 11, even before that, was to extend—

DR. KRAFT:
I think that is an extremely important point. We didn't do anything, operationally from then on, or we did a lot of E&D, engineering and development work, because we had to expand the

envelope of the machine to make it heavier, to do more things and carry more weight and bring more back and everything you can name about it. But it was all done because of wanting to do more science.

MR. MORRIS:
That's right.

DR. MUELLER:
One point to make, and that is the Apollo, as it was conceived and designed, had some limitations on what it could do. Perhaps one of the chief things that we failed to find in our various explorations of the Moon was where water was because that is, of course, the key to future use of the Moon. And that is when we simply didn't have the envelope to simply explore where the most likely places were for the water to be,

which is by the poles. But then we were faced with the fact that we had an envelope that we could carry on scientific research. And beyond a certain point, we weren't able to do enough more things to make it worthwhile to argue at least that you should continue.

Now, in retrospect, the scientists said well, we should have had another 20 missions because we now know what we could do with that fairly limited capability. But by that time, it was past the point of no return. We also had started on, we thought, a rather ambitious program to build a space shuttle, a completely reusable vehicle, and we were going to build a transportation system that was going to be low cost to a space station that was going to serve as a way-point to lunar operations, a real lunar operation. So we had a program in place that

Dr. Robert R. Gilruth, director, NASA Manned Spacecraft Center, with the Snoopy poster, May 11, 1969. (Houston Chronicle Photo, May 11, 1969.)

we thought was going to transform interplanetary space into the kind of thing where people were going to live.

MR. TINDALL:
We had Skylab coming up, too.

DR. MUELLER:
We had Skylab.

MR. TINDALL:
We had a lot of science on that.

DR. MUELLER:
Right.

DR. LOGSDON:
It may be too early in our time to ask this question, but it is the clear follow-up to Dr. Mueller's observation. Why didn't Apollo provide a convincing demonstration to continue that kind of program, that kind of development, to build on it, to continue to use the Saturn V systems to build a large space station, a 33-foot station, to get on with the Shuttle, with lunar bases? Somehow, there was a sense that the country didn't become convinced by what Apollo accomplished that it was in its interest to continue. Was there something in the program itself that led to that outcome?

DR. MUELLER:
I would offer the observation, that it was rather the perception of the President at the time and his set of values that led to a decision both to abandon the Saturn V, because we had never really planned on abandoning the Saturn V at the time we did since we wanted the capability of continuing. For example, we had two Skylabs built and planned to fly both of them. But, the priority shifted. We had to solve the problems of New York City and Vietnam simultaneously, and his priority said that that was a better way to spend the nation's resources.

DR. KRAFT:
Well, I want to give you a humorous answer to that. If George Mueller knew the answer to that, and not let us know how he could have solved the problem, and he didn't know it at the time, I would have been pretty ticked off about it. I don't think there is any answer to that question. I mean, of all people, George Mueller would have given us an answer to it. We just didn't have one. That is all there is to it.

DR. LOGSDON:
But I want you to reflect on the program itself. Why did it become politically acceptable for Richard Nixon to make that decision not to push on with an ambitious program or the program that had been laid out. Was there something about Apollo that did not convince the country to go forward?

DR. GILRUTH:
Well, it was bilateral. The last landing we made on the Moon, the networks wouldn't pay for putting it on television. We had to pay for that ourselves. Did you know that? We had been so successful that people thought they knew all about it and they did not think that they wanted to watch it.

DR. KRAFT:
Well, after Apollo 11, we had two men on the Moon and one circling, and we had real time pictures of it. And I would turn on the three major networks at mission control, and they, all three, had soap operas on. What can we do about that? I don't know.

DR. LOGSDON:
Well, is there something that says that doing space programs as TV spectaculars is not the right way to go about it? I mean, is that the right criteria?

DR. KRAFT:
No, look. Let me tell you. If landing men on the Moon and bringing them back safely, and all of

the exciting things that we did on the lunar surface doesn't excite the country and doesn't want to get them moving, I don't know what the hell will. And I don't know where it is today. That is just the way this country is, I suppose. We all have to recognize that.

DR. LOGSDON:
So how do you do a program that lasts for a long time?

DR. MUELLER:
Well, let me offer an observation. The Russians have done that. As long as we had that commitment on Apollo, we were doing very well indeed. It was when we decided that we had finished that task, the President had decided we had finished it, that we began to lose support of the Congress and of the President and so that withered away. What you need is a commitment, a national commitment to a continuing program that doesn't depend upon a spectacular success, but depends upon some results in the economy.

DR. LOGSDON:
But let me take that one step further and ask the people that ran Apollo. Could you capture the spirit, the elan, the excitement of that 1961 to 1972 period, of that program, with a continuing multidecade program of humans in space? Was there something unique about the goal, the timetable, the ability to put a flight operations team together of young Turks that really went to the job, that can't be reproduced in a more normal environment? Was Apollo special in ways that can't be reproduced?

DR. GILRUTH:
Well, we landed six times on the Moon and we found some differences in the different places. But it was really no point in making more landings to find out more about the Moon. And with that much risk, we had more spacecraft that we could have flown. I didn't want to fly them. I didn't want to send any more people to the

Moon because I thought we had learned the things we were going to, and it was not worth the risk.

DR. LOGSDON:
How about your flight teams? Were they as good on Apollo 17? Were they better? Were they sharper?

DR. KRAFT:
Every one got better.

MR. MORRIS:
Oh, yes.

DR. KRAFT:
They all got better. Let me try and answer your question a little more directly. How you can excite an organization or a nation like we did in Apollo again is, in my mind, extremely difficult to do. Here is a Moon that has been sitting there as long as man has been able to look at it. We have been writing stories about it for hundreds of years. Not tens of years, but hundreds of years. And then we suddenly do it.

Now, to try to come up with an event which is going to recapture the imagination of the United States and the world as comparable to that I don't think is possible. Now, maybe it is. But I don't think it would be for as long. I mean, even if we said we were going to go to Mars, I don't think you could keep that kind of momentum going again.

So, I think George is absolutely right. We have got to take a different tactic which says look, space is extremely important to the economic structure of our country. It is just as important as defense. It is just as important as education. It is an integral part of what we have got to do in this country to remain preeminent, competitive, technologically ahead. And that doesn't say only the space program, but it is one of the important elements. And the words I have used is, it isn't some esoteric fantasy that we bunch

Dr. Christopher C. Kraft director, Johnson Space Center, after the unveiling of the L.B. Johnson bust, reads the citation to Mrs. L.B. Johnson during the 1973 dedication services for naming the Spacecraft Center after the former president. (NASA Photo 73-H-853.)

of space cadets want to do. It is something that is fundamental to the economic structure of our country.

MR. TINDALL:

Do we want to accept the assumption that the people of the United States don't support this? I don't think that is even true. I don't think they have supported it a lot. The business of whether they want to watch it on TV all the time, you start talking about a program that lasts 10 years. I think that if you had a trouble-free program that was flying along marvelously for 10 years, they would get terribly bored with it, particularly compared to the soap operas because, you know, really exciting stuff every-day.

DR. LOGSDON:

Let me turn the discussion inwards. In terms of the excellence of performance of an organization, like NASA overall, like the Johnson Space Center, how do you keep the quality of performance that made Apollo possible, sustained over a period of decades without that kind of challenging goal, that short-term excitement?

DR. MUELLER:

I would argue that that really is quite possible within an organization. And I will use an example. Bell Laboratories had no over-reaching goal, but it has had some very dedicated people working in an environment, which was very conducive to forward-looking work. The NACA didn't have an over-reaching goal, but it had some very dedicated people who worked continually and did some really astonishing things. Our problem in the space arena has been that we have first fluctuated our resources that we have applied, and we have failed to provide a vision of what it is we were trying to accomplish in any fashion, really. And that lack of a feeling of where you are going and the inability to hire at the right level the right set of people on a continuing basis has been a problem. But I don't think there is any difficulty in keeping an organization motivated if it has a place to go with the things it is going to accomplish.

The public perception of that can be quite different. And keeping Congress motivated is a better question, or how you keep the OMB motivated. That is where the real question lies.

DR. LOGSDON:
Anybody else want to comment on that issue?

DR. FAGET:
Well, I certainly agree with George. My memories of the NACA, there were very few people that weren't very motivated, that didn't do a great job. The NACA shunned publicity, literally shunned publicity. And 95 percent of their programs were in a classified nature; they couldn't talk about them anyway. But even their unclassified programs, they didn't go out of their way to brag about what they were doing. People knew they were doing good work and they were motivated.

DR. KRAFT:
I will give you a different answer. I would like to take the six people sitting here, I would like for you to give me maybe a couple hundred more, I would like you to give me a budget, and I would like you to send me off to do it again.

DR. LOGSDON:
And you think you are ready?

DR. KRAFT:
But I think, I am being somewhat cynical, but I think that it would take something like that to get it started again. I think that this agency has become so bureaucratic in the last 10 years, and particularly accelerated since the Challenger accident, that trying to reinitiate this organization into something is going to be very difficult to do. It "ain't" going to be easy.

I often made speeches around this center here and said look, I don't know what it is, I don't know what the glue is that makes this organization as great as it is. But let me tell you something, if we ever lose it, I won't know how to put it back together again. And I am frightened to death of that. And so I think we are rapidly approaching that in NASA, and I am frightened to death of it.

DR. LOGSDON:
I would like to hear the rest of you react to that. What is it about the organization today? We are sitting here July 21, 1989. Yesterday, President Bush said we are going back to the Moon, we are going on to Mars. He didn't put a timetable on it. What if he had done that? What if he challenged the agency in 1989 to do another major human program beyond Earth orbit?

DR. KRAFT:
If he had written a check, I would have volunteered.

DR. LOGSDON:
So it all comes down to money?

DR. KRAFT:
But he hasn't written a check.

MR. MORRIS:
It is more than money. It is motivation. The check is a form of commitment, that I really want that job done. If you have that kind of commitment from the leadership, I don't think there is any doubt that you can get enough good people to go make the thing happen again.

As a matter of fact, we are kind of talking as if the situation deteriorated so that there are not good people doing good things. I don't think that is true. The people I go talk to here at the center and at Headquarters and at the Cape and other places, I find guys working just as hard now as we did back in the 60s and early 70s. I really do.

DR. LOGSDON:
What's the difference, then, Owen?

MR. MORRIS:
The difference is the commitment from above, to some extent. I don't think we have had the leadership within the Agency and the relationship between the Agency and the administration for the last 20 years that we had in the first 10—

MR. TINDALL:
It is not just dollars.

MR. MORRIS:
—I think that is a big part of it.

MR. TINDALL:
It is not just dollars. Everyone knows that it is having a schedule, a certain deadline that you have to meet.

MR. MORRIS:
Right. Exactly.

MR. TINDALL:
If you are going to launch on a certain date, that gets everybody's attention. And everyone knows that they have got their job to get done by then or else they are the ones that are holding it up.

DR. GILRUTH:
God forbid, that schedule pressure.

MR. TINDALL:
It is schedule pressure. You've got to understand, where I am working now, there isn't any schedule pressure at all. Some projects are going along, and I am just astounded, all of a sudden a project manager will have a schedule slip of a year and not care. I am not kidding you.

DR. KRAFT:
I will go even further than that. I have spoken to one of the major program managers in the last few days and said we need to get on with

the space station. And his answer was well, you may want it in a hurry, but what is the hurry.

MR. TINDALL:
Yes.

DR. KRAFT:
What difference does it make whether we get the space station in two or three years. And frankly, I think that is an attitude that does exist. And if you have got that attitude, you ain't going to get it done.

MR. TINDALL:
Because all of the other organizations have to know that there is a certain launch date, and whatever it is going to be they are going to be held accountable if they are not ready. Really puts a lot of motivation.

DR. LOGSDON:
How do you do this decades-long thing?

DR. KRAFT:
I don't know. Maybe there is some serum you give somebody. I don't know, but I said that is a very tenuous thing. There isn't any formula that you can write down to make that happen.

DR. MUELLER:
It's called leadership. And fundamentally, unless you have got the right set of leaders in, you can't cause it to happen. If you have got the wrong set in, it won't ever happen.

DR. GILRUTH:
Yes, but you have to have some real knowledge that you can get the funds and you can do it if you put together a good enough group. Otherwise, you are like we are now. We can't put a good bunch together because we don't know where we would get the money.

DR. LOGSDON:
Let me go around, maybe start with Chris, and go around, the main audience for this videotape

is the young engineers of the '90s and beyond. What would you tell them the lesson of Apollo for them is?

DR. KRAFT:
Well, unfortunately, there are a lot of them. But you start with the fact that the legacy of Apollo is you can do anything you set your mind to do in this country. And as evidenced by Apollo, it didn't even help to be very reasonable because there weren't very many of us who thought Apollo was reasonable the day that Mr. Kennedy said we were going to do it. We didn't even have a man in orbit yet.

The second thing is, that I think that you can't be afraid of hard work. Because God knows that there is not anybody that worked at the Johnson Space Center, in the '60s particularly, that didn't have 10 jobs to do and they had to have them done tomorrow, and they spent all of their waking hours doing it. Even when they were at home having dinner at night they were thinking about getting their job done, and enjoyed every damn minute of it. And I would like to be faced with that problem again.

DR. LOGSDON:
Bob.

DR. GILRUTH:
I think it has been said pretty well, and I don't have anything to add right now.

DR. LOGSDON:
Max.

DR. FAGET:
Well, of course Apollo was really a unique situation, and there is no way that we can recreate that. But, basically, I really think that a lot could be done towards getting the engineers in NASA, giving them the opportunity to do more on their own behalf. I know that going back to the good old days, to the NACA, is a song that everybody is tired of hearing, but back in those days, it didn't make any difference how junior an engineer you were. If you showed any spark of capability, you could bet that your supervisor would give you a project that would be your own, your own responsibility. And you would be entirely graded by what you did based on your own thinking out of the problem and accomplishing it. And that is what you need.

Howard W. Tindall, Jr., second from right, and, left to right, Bill Schneider, Chris Kraft, and Sig Sjoberg monitor a problem with the Command Service Module used to transport the Skylab 3 crew to the orbiting Skylab space station cluster. (NASA Photo S-73-31875.)

You need an organization with a culture in which an individual can take on a job and get it done if given a chance. And I am afraid that has to be instilled and I agree with Chris, you need leadership to do that. The leader has got to really believe in his organization, and believe that they can do things, and find ways to challenge them.

DR. LOGSDON:
Owen.

MR. MORRIS:
I think one of the big things that I saw in all of the people working on Apollo was a sense of pride in doing something right. A sense of being willing to take a challenge, find out how to go about it, and then go solve it, make the event happen whatever it was, and then the sense of pride that comes from that. And there is willingness there to go beyond what you are asked to do, to go get your hands dirty, to go understand how the hardware is really going to work, to understand how the software is really going to interface, whatever the job is. And I think that willingness to go the extra mile, to get the feeling of satisfaction you get when you do the job right is something that is really important then.

DR. LOGSDON:
George?

DR. MUELLER:
Well, I had a somewhat different career, in a sense, because my first 20 years as an engineer and physicist were involved in avoiding management. So, I spent a lot of time working hard on benches and things like that and teaching, and trying to keep from becoming a manager. And I think that is something I would like to pass on to the young people coming in the years ahead, and that is, until you really understand and have built within yourself the technical capability of knowing something very well and in-depth, you really ought not to try to man-

age a program. And so often we think that management is the end point. It really is just one of those things that you have to do as career develops, but it isn't the "funness" part at all. The fun part is when you are doing something. And that is what we fail to really emphasize in today's world.

MR. MORRIS:
Amen.

DR. LOGSDON:
Bill.

MR. TINDALL:
Well, I don't have that much to add. I think both the thing that Chris said and that George said is exactly right. I remember through my career, I never worried about the next, or any promotions at all. In fact, I was just having a really good time, really, really good time. And I guess the organization we were in encouraged that. That is what both of them were saying. You didn't really, you weren't concerned about whether you were going to become a program manager and quit lifting those 50 pound weights up and down hundreds of times a day like I used to do.

DR. KRAFT:
Or have hydraulic oil spilled all over you.

MR. TINDALL:
That's exactly right. We don't want to get in there. But the thing that was so outstanding, you just hope that the young engineers and scientists that we are talking to here have a chance to be, to get into an organization, I don't know whether it has to be a project like Apollo, but an organization like we had that really delegated the jobs as tough or tougher than you could do and just said go on out there and figure out how you can do it because it was so doggone much fun.

I mean, when Chris was talking, when he started out talking about these terrible 14 and 16 hour days and all during dinner that you were working, but I would just change the word from work to play because I never thought we were working at all. And that is the honest to God truth. It was just so much fun. In fact, I think it would be terrible if you had to go through life working. Really.

DR. MUELLER:

I agree. I would like to add one thing for our future managers. Just because you become a manager doesn't mean you no longer can do anything technical. Some managers think that you are a manager, and therefore you can't do anything useful anymore.

I remember sometime, about 1968 I guess it was, that Sam Phillips asked me to take off my hat of manager and start looking at software and try to do an in-depth review of our software system. It happens that I have some back-ground in software systems. So I did do that. So that was quite a separate thing from being manager of a manned spaceflight program. It was a technical challenge, and we did a fairly in-depth review and I think we had some positive impact as a result of that in terms of being sure that the software wasn't going to destroy us half way where ever we were.

DR. LOGSDON:

Apollo, 20 years ago yesterday, for the first time landed humans on the Moon. It is clearly a piece of history, a grand human adventure, but it was also a remarkable engineering, technical, management operation achievement. The six gentlemen that have been sharing their thoughts with you this afternoon are the people, and represent many more people, that made that happen. And we may not see the likes of Apollo again, but the future will hold great promise as well.

About the
Moderator

John M. Logsdon is director of both the Center for International Science and Technology Policy and the Space Policy Institute of George Washington University's Elliott School of International Affairs, where he is also professor of political science and international affairs. He holds a B.S. in physics from Xavier University and a Ph.D. in political science from New York University. He has been at George Washington University since 1970, and previously taught at The Catholic University of America. He is also a faculty member of the International Space University and Director of the District of Columbia Space Grant Consortium.

Dr. Logsdon's research interests include U.S. and international space policy, the history of the U.S. space program, and the structure and process of government decision-making for research and development programs, and international science and technology policy. He is author of *The Decision to Go to the Moon: Project Apollo and the National Interest*, general editor of the highly acclaimed series *Exploring the Unknown: Selected Documents in the History of the U.S. Civil Space Program*, and has written numerous articles and reports on space policy, space history, and science and technology policy. He is North American editor for the international journal *Space Policy*.

He is an elected member of the International Academy of Astronautics and the Board of Trustees of the International Space University, and Chair of the Advisory Council of the Planetary Society. He is a fellow of the American Association for the Advancement of Science and the American Institute of Aeronautics. He is currently a member of the Committee on Human Exploration of the Space Studies Board, National Academy of Sciences, the Commercial Space Transportation Advisory Committee of the Department of Transportation, and on a blue-ribbon international committee evaluating Japan's National

Space Development Agency. In past years, he was a member of the Vice President's Space Policy Advisory Board, the Aeronautics and Space Engineering Board of the National Research Council, the National Academy of Sciences—National Academy of Engineering Committee on Space Policy and the NRC Committee on a Commercially Developed Space Facility, NASA's Space and Earth Science Advisory Committee, and the History Advisory Committee of the National Air and Space Museum. He is a former chairman of the Committee on Science and Public Policy of the American Association for the Advancement of Science and of the Education Committee of the International Astronautical Federation. Dr. Logsdon has lectured and spoken to a wide variety of audiences at professional meetings and colleges and universities, international conferences, and other settings, and has testified before Congress on numerous occasions. He is frequently consulted by the electronic and print media for his views on various space issues. He has been a fellow at the Woodrow Wilson International Center for Scholars and was the first holder of the Chair in Space History of the National Air and Space Museum. Dr. Logsdon has served as a consultant to many public and private organizations.

Monographs in Aerospace History

Launius, Roger D., and Gillette, Aaron K. Compilers. *The Space Shuttle: An Annotated Bibliography.* (No. 1, 1992).

Launius, Roger D., and Hunley, J.D. Compilers. *An Annotated Bibliography of the Apollo Program.* (No. 2, 1994).

Launius, Roger D. *Apollo: A Retrospective Analysis.* (No. 3, 1994).

Hansen, James R. *Enchanted Rendezvous: John C. Houbolt and the Genesis of the Lunar-Orbit Rendezvous Concept.* (No. 4, 1995).

Gorn, Michael H. Hugh L. *Dryden's Career in Aviation and Space.* (No. 5, 1996).

Powers, Sheryll Goecke. *Women in Aeronautical Engineering at the Dryden Flight Research Center, 1946-1994* (No. 6, 1997).

Portree, David S.F. and Trevino, Robert C. Compilers. *Walking to Olympus: A Chronology of Extravehicular Activity (EVA).* (No. 7, 1997).

Logsdon, John M. Moderator. *The Legislative Origins of the National Aeronautics and Space Act of 1958: Proceedings of an Oral History Workshop* (No. 8, 1998).

Rumerman, Judy A. Compiler. *U.S. Human Spaceflights: A Record of Achievement, 1961-1998* (No. 9, 1998).

Portree, David S.F. *NASA's Origins and the Dawn of the Space Age* (No. 10, 1998).

Logsdon, John M. *Together in Orbit: The Origins of International Cooperation in the Space Station Program* (No. 11, 1998).

Phillips, W. Hewitt. *Journey in Aeronautical Research: A Career at NASA Langley Research Center* (No. 12, 1998).

Braslow, Albert. *Laminar Flow Control Flight Research at Dryden, from the 1960s to the F-16XL* (No. 13, 1999).

Those monographs still in print are available free of charge from the NASA History Division, Code ZH, NASA Headquarters, Washington, DC 20546. Please enclosed a self-addressed, 9x12" envelope stamped for 15 ounces for these items.

Index

www.ingramcontent.com/pod-product-compliance
Lightning Source LLC
Chambersburg PA
CBHW081620170526
45166CB00009B/3046